Alan McKirdy has written many popular books and book chapters on geology and related topics and has helped to promote the study of environmental geology in Scotland. His other books with Birlinn include *Set in Stone: The Geology and Landscapes of Scotland* and *Land of Mountain and Flood*, which was nominated for the Saltire Research Book of the Year prize. *Northern Highlands – Landscapes in Stone* was long-listed for the Highland Book Prize. He is also the author of *James Hutton: The Founder of Modern Geology*. Before his retirement, Alan was Head of Knowledge and Information Management at Scottish Natural Heritage. Alan is now a freelance writer and has given many talks on Scottish geology and landscapes at book festivals and other events across the country.

Southern Scotland

LANDSCAPES IN STONE

Alan McKirdy

For Alistair Moffat

First published in Great Britain in 2022 by
Birlinn Ltd
West Newington House
10 Newington Road
Edinburgh
EH9 1QS

www.birlinn.co.uk

ISBN: 978 178027 748 6

Copyright © Alan McKirdy 2022

The right of Alan McKirdy to be identified as the
author of this work has been asserted by him in accordance
with the Copyright, Designs and Patents Act, 1988

All rights reserved. No part of this publication may
be reproduced, stored, or transmitted in any form, or
by any means, electronic, mechanical or photocopying,
recording or otherwise, without the express written
permission of the publisher.

British Library Cataloguing-in-Publication Data
A catalogue record for this book is available
on request from the British Library

Designed and typeset by Mark Blackadder

FRONTISPIECE:
Mudflats at Caerlaverock National Nature Reserve.

Birlinn Ltd would like to thank

for their generous donation towards this publication.

Printed and bound by Gutenberg Press Ltd, Malta

Contents

	Introduction	7
	Southern Scotland through time	8
	Geological map	10
1.	Time and motion	11
2.	Early life of the Iapetus Ocean	14
3.	Death of an ocean	18
4.	Old Red Sandstone times	21
5.	The Study of geology started right here!	22
6.	Volcanic interludes	24
7.	Meandering rivers and shallow seas	27
8.	Coal swamps	29
9.	Desert storm	30
10.	Mineral treasures	32
11.	The Ice Age and beyond	33
12.	The coast	38
13.	Human imprint on the landscape	40
14.	Places to visit	44
	Acknowledgements and picture credits	48

Introduction

If we peer deep into our geological history, we find Southern Scotland has a long and turbulent past. Perhaps most notably it marks the place where an ocean, once as wide as the North Atlantic, was compressed by a convergence of ancient lands and then ceased to be. In time, what rose from this collision zone are the rounded hills we now recognise as the Southern Uplands. In the white heat of continent-to-continent collision, granites formed in the depths of the Earth's crust and ascended close to the surface. Later events exposed them as the most imposing landscape features of Galloway.

It is also the place where the modern science of geology was born. James Hutton, star of the eighteenth-century Scottish Enlightenment, found inspiration from his study of the rocks at Siccar Point on the Berwickshire coast. This place is still hailed for its rock exposures, which are amongst the most historic and important to be found anywhere in the world.

Deserts covered the land that would become Southern Scotland. Thick layers of brick-red-coloured rocks, known as the Old Red Sandstone, piled up, having been dumped by ephemeral rivers and streams that flowed across the parched landscape. It was also a time of violent explosive volcanic activity. This gave rise to the prominent landscape features recognised today as the Eildon Hills. These eruptions continued for another 50 million years. Thick sequences of basalt lava were belched from a myriad of small volcanic vents around Kelso in what is now the Tweed Valley. To the south, the Cheviot volcano had just blown its top. Warm, shallow seas fringed by low-lying floodplains covered much of the land where sandstones and limestones gradually accumulated. Some of these rocks preserve an amazing lode of fossils that tell of a rich diversity of ancient fish species, king crabs, arthropods and predatory shrimps.

In later geological times, the area was covered by a desert like that of the Sahara today. This book tells the tale of these momentous events, written in stone.

Opposite.
View of Moffatdale.

Southern Scotland through time

Period of geological time	Millions of years ago	Scotland's global position	Environments and events in Southern Scotland
Anthropocene	Last 10,000 years	57° N	The first evidence of our species *Homo sapiens* is found in the peat bogs where the pollen signature changes, around 5,000 year ago, when people first inhabited the area and reduced the tree cover to plant crops.
Quaternary	Started 2 million years ago	Present position of 57° N	• **11,500 onwards** – the ice retreated as the climate started to warm. • **12,500 to 11,500 years ago** – the climate became very cold as the ice returned. **14,700 to 12,500 years ago** – for a brief interlude temperatures were similar to those of today. • **29,000 to 14,700 years ago** – the landscape was entirely covered by an icesheet during this, the last advance of the ice. • **Before 29,000 years ago** and for a period approaching the last 2 million years, there were prolonged periods when thick sheets of ice covered the area. These advances of the ice were separated by warmer interludes, known as inter-glacials, when the temperatures rose to levels similar to those of today.
Neogene	23–2	55° N	Temperatures fell as the Ice Age approached.
Palaeogene	66–23	50° N	Between 65 and 60 million years ago, the ancient continent of Pangaea was split asunder and the North Atlantic Ocean began to form.
Cretaceous	145–66	40° N	Sea levels rose to cover the area, but no rocks of this age are preserved in Southern Scotland.
Jurassic	201–145	35° N	No deposits of this age found in the area.

Period of geological time	Millions of years ago	Scotland's global position	Environments and events in Southern Scotland
Triassic	252–201	30° N	Desert conditions prevailed across what is now Scotland. Small deposits of Triassic strata are to be found in the eastern part of Dumfriesshire.
Permian	299–252	20° N	Desert conditions were widespread and the remains of these dunes are to be found near Dumfries, Lochmaben and Thornhill.
Carboniferous	359–299	On the Equator	'Scotland' was located at the Equator at this time. Tropical swamps were widespread.
Devonian	419–359	10° S	Desert conditions prevailed throughout this time. Ephemeral rivers deposited sand and conglomerates. Explosive volcanoes were also erupting at this time.
Silurian	444–419	15° S	Large upheavals created the Highlands of Scotland and the Southern Uplands, as the Iapetus Ocean closed.
Ordovician	485–444	20° S	'Scotland' was located on the shore of the Iapetus Ocean, which had started to close by this period. Thick deposits of sandstones and shales (known collectively as turbidites) were deposited during this period. Some volcanic activity also occured.
Cambrian	541–485	30° S	'Scotland' and 'England' were separated by the width of the Iapetus Ocean. No rocks of this age have been recorded in the area.
Proterozoic	2,500–541	Close to the South Pole	No rocks of this age are preserved in Southern 'Scotland'.
Archaean	Prior to 2,500	Possibly close to the South Pole	The age of the Earth is around 4,540 million years.

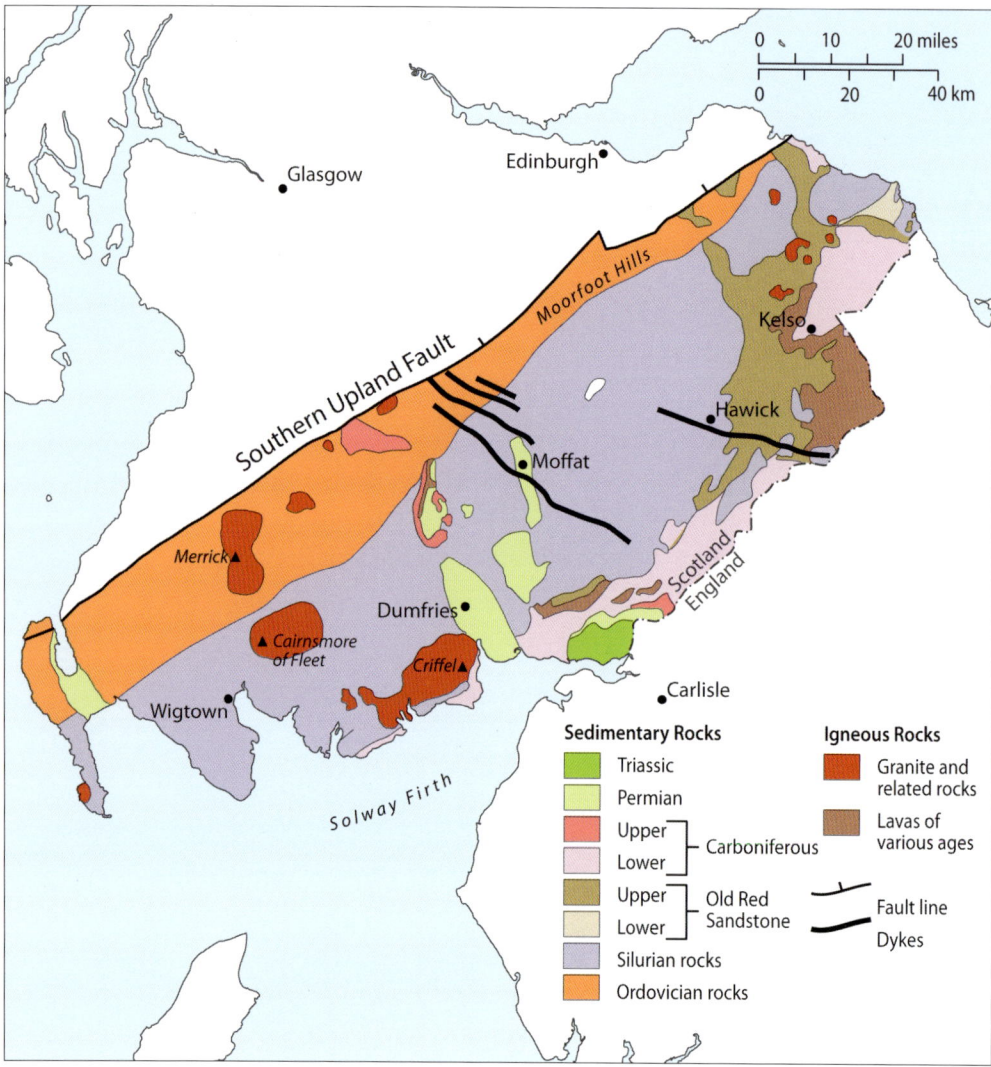

The geological map of Southern Scotland is a patchwork of rocks of different types and ages. Dominating are the rocks of Ordovician and Silurian age that were laid down as sands and muds in an ancient and now long disappeared ocean. The Iapetus Ocean existed for around 250 million years but disappeared as the continents were rearranged and subsequently collided. Granites, once molten reservoirs of magma, were generated during this coming together of continental landmasses. This was an unstable place after the collision and volcanoes were active across the area for around 50 million years. Then a period of calm returned, particularly in the east of the area, where older rocks of the Southern Uplands were buried by a veneer of younger strata rich in fossils. Desert conditions returned during Permian and Triassic times, with deposits of useful building stones preserved around Dumfries, Lochmaben and Stranraer. Thin ribbons of dolerite were shot across the area, as volcanoes of the Inner Hebrides (Mull particularly) were active. A final burnishing and reshaping of the area occurred during the Ice Age, when the area was submerged by a thick carpet of ice and snow.

1
Time and motion

Time

The timescale over which the events that built Southern Scotland were played out is unfamiliar almost to the point of being unfathomable. We may consider the ancient Pictish standing stones and abbey ruins of the area to be the furthest back in time the area can take us, but we must rewind many more thousands of years – in fact, millions of years – to find the truth of it. The Earth is 4,543,000,000 years old (or 4.543 billion years) and this part of the world can claim a geological history that stretches back around 500,000,000 (500 million) years. For many readers, these numbers will be difficult to grasp, but grasp them we must if the unfolding geological history of the area is to make any kind of sense. James Hutton, who spent 14 years of his life farming near Duns in the eighteenth century, used local places to unlock the timescales involved. But it would be two centuries before specific numbers could be applied. In studying these strata, Hutton could find 'no vestige of a beginning – no prospect of an end', in terms of the age of the Earth and the timescale required to form the rock structures he observed. His words have resonated down the ages and heralded the birth of a new science – modern geology. This and many of Hutton's other deductions are as true today as they were when he published his great works over 200 years ago. We'll return to Hutton later.

Geologists have divided these yawning stretches of time into manageable chunks known as 'periods'. This approach helps scientists to correlate rocks and events of similar age across the country and indeed around the world. Pages 8 and 9 of this book place the events that have left a mark on this area in date order. This sequence of events is known as a geological column. Telling the story of events and changing environments in this abbreviated manner is one of the cornerstones of geological science.

Motion

Another difficult concept to appreciate is that the ground beneath our feet is moving. In fact, the continents have been relentlessly travelling across the globe since the dawn of time. The Earth's core is as hot as the surface of the Sun at around 6,000°C and, inevitably, heat leaks upwards and reaches our realm, the surface of the planet.

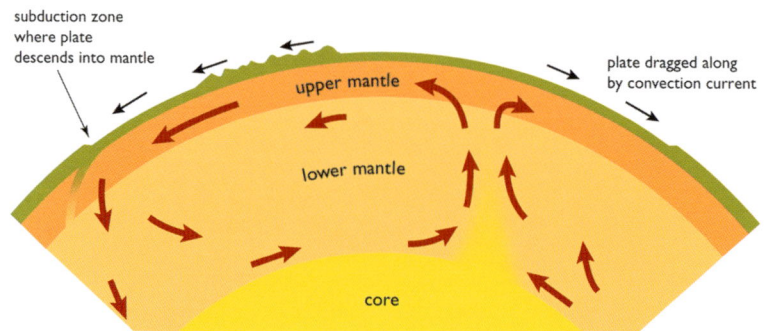

Heat radiates from the Earth's core and sets up a convection motion. These currents of hot rock circulate beneath the upper layer of the planet – the crust – and drive the continents across the face of the globe. They don't travel at a fast pace – on average around 6 centimetres per year – but over millions of years continents can move from one side of the Earth to the other.

The Earth's surface is divided into a series of seven large plates and around a dozen smaller ones. They are propelled by forces under the Earth's crust, created by the circulation of hot rocks beneath, and each moves independently across the surface. They grind past or under each other and occasionally continents collide to create mountain ranges. This concept is known as plate tectonics.

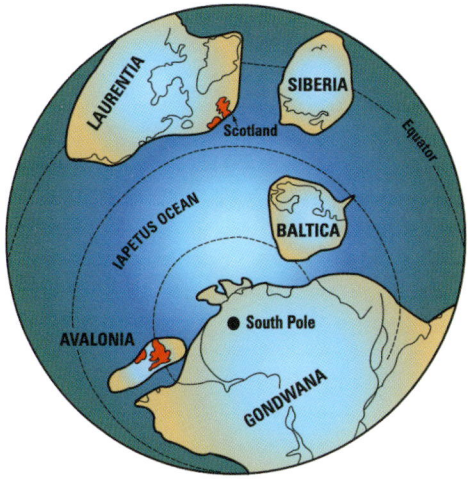

a. Late Cambrian c.500 million years ago

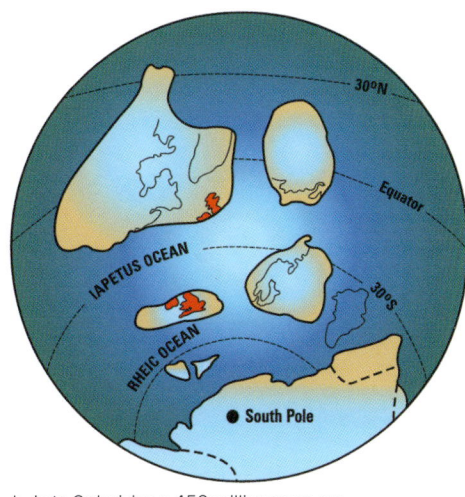

b. Late Ordovician c.450 million years ago

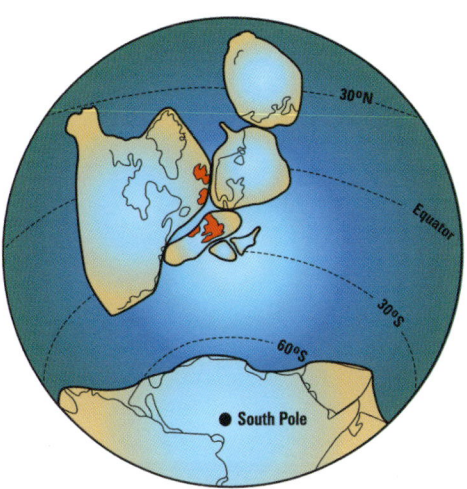

c. Early Devonian c.400 million years ago

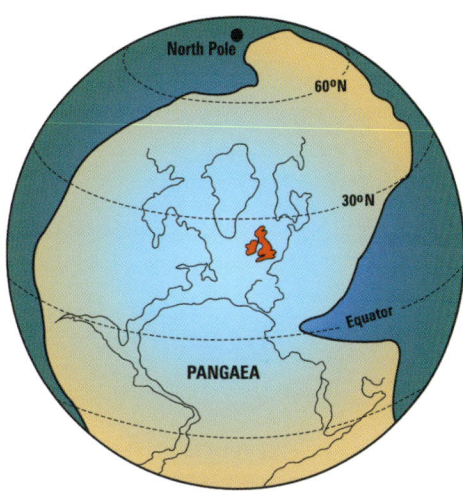

d. Permian/Triassic c.250 million years ago

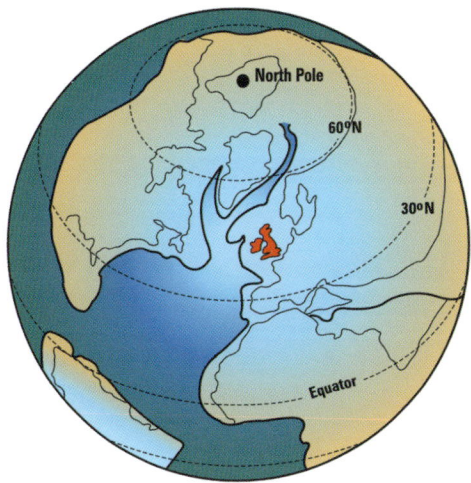

e. Palaeogene c.50 million years ago

This series of globes charts the changing position through time of the land that would become Scotland. Around 500 million years ago, two parts of what is now the United Kingdom were separated by an ocean wider than the present-day North Atlantic. This expanse of water was known as the Iapetus Ocean. Over time, Scotland, which was originally 'a little piece of North America', travelled closer to England and Wales. Eventually, around 420 million years ago, the Iapatus Ocean closed and continents collided. During the next 300 million years, a supercontinent known as Pangaea (or 'All Earth') was formed, as continents coalesced. Drifting ever-northwards, driven by forces deep in the Earth, in another re-arrangement of the continents, the North Atlantic Ocean formed and Scotland travelled as part of what would become continental Europe.

2
Early life of the Iapetus Ocean

Today, we are comfortable with the reassuringly familiar distribution of land and sea across the face of the planet. However, the more challenging continental associations and ancient oceans of past geological times, such as those illustrated in the reconstructions on p. 13, must be embraced if we are to understand how the geological story of Southern Scotland unfolded.

Around 500 million years ago, Scotland was a small part of a continental landmass known as Laurentia. It was separated from what would become England and Wales by the width of the Iapetus Ocean, a stretch of water many thousands of kilometres wide. Baltica, later Scandinavia, and Siberia lurked offshore from Laurentia. It's heady stuff, and as geologists our 'big picture' understanding of the early development of Scotland is based on this interpretation.

The continents that bounded this great ocean were the source of the boulders, sand, silt and muds that were carried down by rivers and streams from the high ground to the sea. These flowing waters dumped their burden of sediments close to shore in the shallows. The layers of sands, silt and mud built up on the continental shelf over the years and, from time to time, great avalanches tumbled into deeper water in a single catastrophic event. These are known as turbidity flows and the resultant deposits that spread across the ocean floor are described as turbidites.

These turbidity flows race down the slope of the continental margin and subsequently travel for hundreds of kilometres across the ocean floor, building up in thick layers of sediments.

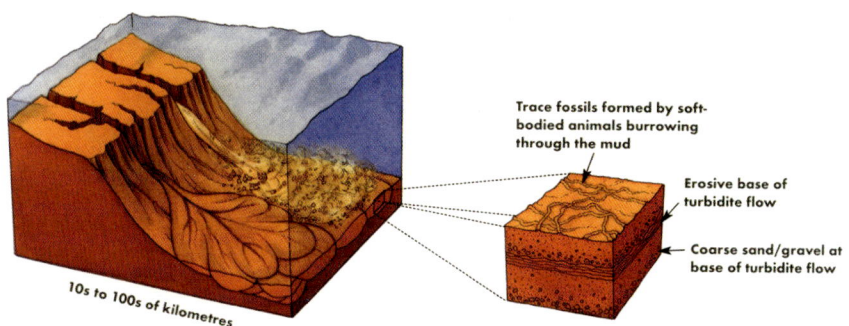

The bulk of the material that comprised the initial flow was sand and some larger stones, and then a layer of mud settled out on top over a much longer period. This process of build-up in shallow water and then of chaotic collapse into the ocean below was repeated many millions of times during the Ordovician and Silurian Periods. This equates to a time interval from around 485 to 419 million years ago. These turbidite flows have a rhythm of rock layers that is characteristic and readily identifiable in the geological record. Thick layers of sand alternate with thinner muds in an endlessly repeating pattern.

The monotonous repetition of the sand and mud layers that comprised the thick sequence of turbidite layers is leavened by the presence of enigmatic fossils known as graptolites. Today, they appear as pencil-like marks on the rock surface. They lived as free-floating colonies of marine creatures during the period when the Iapetus Ocean existed. These fragile, thread-like animals were incorporated into the build-up of sediment on the sea floor when they died and sank to the seabed, becoming buried by the settling mud. These animals rapidly evolved through the ages. There were many different types, or species,

Regular alternations of layers of sands and mud characterise turbidite deposits and are readily recognisable – like here in Ardwell Bay. They would have originally been laid flat on the ocean floor, but have been buckled and upended by subsequent Earth movements.

Graptolites took many different forms. From the simplest blade-like form, known as monograptus, they varied hugely in shape and geometry. Dicellograptus had two blades, known as stipes, whilst others had a dendritic or branching form. The graptolite featured here is monograptus.

Opposite. This is Lapworth's detailed map of the Dob's Linn area near Moffat. It demonstrates the work that was required to establish the geological structure of the area. He used a detailed knowledge of the graptolites contained in each rock layer to establish their relative position in the geological sequence. It was painstaking and slow work.

of graptolites that co-existed in the oceans of the ancient world. These animals attained worldwide distribution and the fact that they evolved rapidly into many different forms made them ideal fossils to link deposits of the same age around the globe. They also proved to be extremely useful in untangling the complexities of the geological history of the Southern Uplands. This highly successful and widely distributed group of animals were largely creatures of the Lower Palaeozoic and became extinct shortly after the Iapetus Ocean closed.

Working out the sequence of geological events created by these events initially fell to a school teacher named Charles Lapworth (1842–1920). He taught first in Galashiels and then took up a position at Madras College in St Andrews. He was intrigued by the fossils he found near his first school in the Scottish Borders and that launched him into an intensive study of the Southern Uplands starting in 1869. He devised novel methods and techniques to unravel the complex structures he was confronted with and was eventually rewarded with a professorship at an institution that would later become the University of Birmingham.

Map and Sections of the Moffat Rocks of the Typical locality of Dobb's Linn.

3
Death of an ocean

As the continents approached, the layers of mud, sands and limestone that had built up over millions of years on the ocean floor were folded and buckled, as if squeezed in an oversized vice. Mountains were raised to the north, which we now recognise as the Highlands of Scotland. Later, the continents nudged together and the Iapetus Ocean was finally extinguished. The sands and muds laid down in date order from oldest to youngest on the ocean floor were folded and, in many cases, turned up on end to sit in a vertical plane. The full force of continent-to-continent collision was exerted on these layers of sand and mud. The new configuration of the continents delivered by the closure of the Iapetus Ocean is well illustrated in the 'Early Devonian' globe on p. 13.

Continents never stay in the same place for long, as they are driven by the unrelenting currents of hot rock just beneath the Earth's crust. The Iapetus Ocean, which existed for around 250 million years, narrowed and finally closed around 420 million years ago. But that process was anything but rapid. The continents of Laurentia and Avalonia, which carried early versions of Scotland and England respectively, inched together over a 60-million-year period.

The place where these continents collided is still preserved, albeit deeply buried for the most part. It is known as the Iapetus Suture – the join created when Scotland and England came together around 420 million years ago. It follows the trend of the Solway Firth northeastwards towards Holy Isle on the east coast. Its trace is buried by younger rocks of Triassic and Carboniferous age, but using deep-penetrating seismic waves its presence can be detected beneath these younger strata. It seems that during the collision Scotland came out on top!

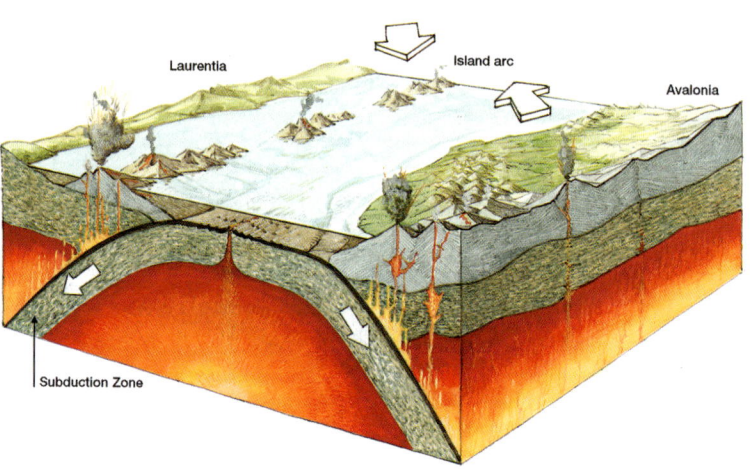

Left.
Evidence of the forces involved are plain to see. These layers of sand and mud in Wigtownshire show the effects of this coming together of ancient continents.

Below.
On the east coast, the effects of this catastrophic event are equally evident. This view of the coastline near St Abb's Head in Berwickshire shows rocks, also from the floor of the Iapetus Ocean, buckled by these earth movements.

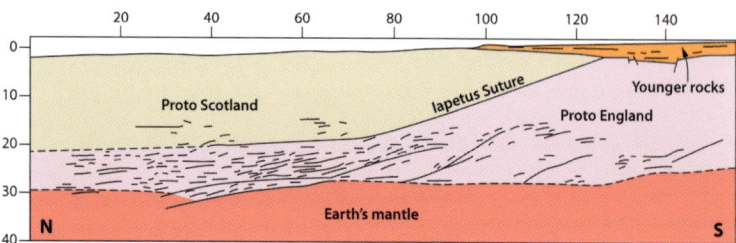

This north–south slice through the Earth's crust shows the outcome of the collision between proto-Scotland and proto-England. As the Iapetus Ocean closed 420 million years ago, the leading edges of each continent locked together for eternity. This geophysical seismic survey reveals that the 'Scottish' crust overrode its 'English' counterpart. The line of the contact between the two is known as the Iapetus Suture. It does not appear at the surface, as it was later covered by younger strata. It is fascinating to note the strong correlation between the line of the Iapetus Suture and the present-day border between Scotland and England.

Granites galore

The closure of the Iapetus Ocean also saw the generation of copious amounts of granite magma. In the white heat of continent-to-continent collision, great balloon-shaped bodies of molten rock formed at depth, later to ascend to a higher point in the Earth's crust. These bubbles of molten granite magma were emplaced close to the Earth's surface, as it existed at the time. The granites only saw daylight when the overlying rocks were subsequently stripped away by erosive forces. They are shown in blood-red colour on the geological map (p. 10). Today we recognise these large masses, which geologists describe as batholiths, as the Loch Doon, Cairnsmore of Fleet and Criffel-Dalbeattie granites.

It is also interesting to note that a major granite body still lurks beneath the surface, perhaps to make an appearance at some future date. Near Abbey St Bathans in Berwickshire, there is a small outcrop of granite. About 10 km below, in the depths of the Southern Uplands, geophysical studies have identified another great mass of granite, tentatively described as the 'Tweeddale Pluton'.

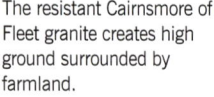

The resistant Cairnsmore of Fleet granite creates high ground surrounded by farmland.

4
Old Red Sandstone times

After the continents locked together, a new world order was created. Geologists call this new construction 'the Old Red Sandstone' continent. It was still an unstable and unpredictable place. Earthquakes were common and explosive volcanic eruptions shattered the peace at irregular intervals. The land that would become the British Isles was still positioned south of the equator on its journey northwards. Conditions were desert-like and the process began of wearing down the mountains created by the continental collision. Great quantities of rubble were generated from these crumbling ruins by the erosive effects of wind and water. Debris piled up on the lower ground, creating thick deposits of red sandstone and conglomerates. We recognise this build-up of the products of erosion as the Old Red Sandstones, characteristic of the Scottish Borders area from Hawick through Jedburgh to Eyemouth.

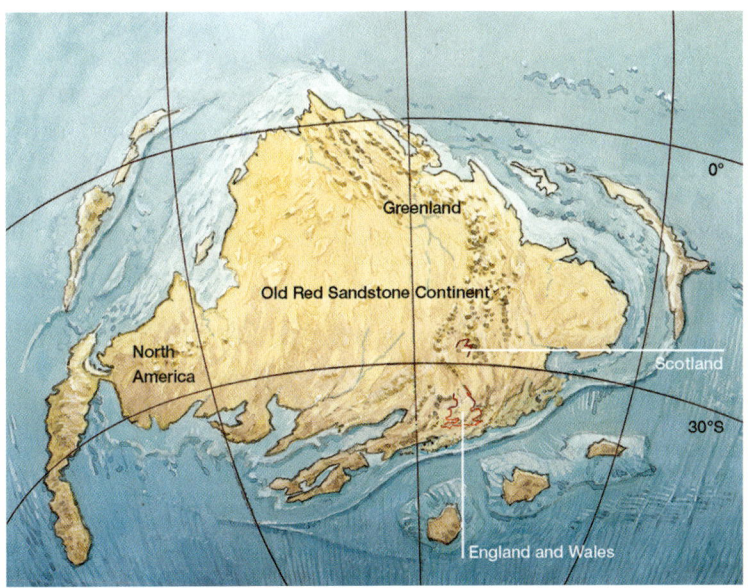

The Old Red Sandstone continent was created after the Iapetus Ocean closed. It comprises the land that would later become North America, Greenland and bits of northern Europe. 'Scotland' had moved significantly northwards but still remained south of the equator.

5
The study of geology started right here!

Today, geology is studied by students all over the world. But the science was founded right here in Berwickshire around 230 years ago. A Borders farmer, James Hutton lived near Duns for 14 years of his life and made his name studying the rocks of his native Scotland and understanding how they formed. During the late eighteenth century, most people believed in a literal interpretation of the Bible – that the world was created in six days and one of rest – but Hutton thought differently, and his evidence was close at hand.

Siccar Point near St Abb's Head and Allars Mill near Jedburgh were two places that gave him cause for thought. The underlying rocks of

The underlying rocks are dated at 450 million years and are overlain by the Old Red Sandstone which are 50 million years younger. Although Hutton was not able to date the rocks, he knew that they were separated by a great gulf of geological time. In his famous book, *Theory of the Earth*, which laid the foundations for modern geology, he said that, looking at these unconformities, he could see 'no vestige of a beginning – no prospect of an end' to geological time. This famous deduction and observation that some rocks were once in a molten state secured for this man the title of Founder of Modern Geology. Siccar Point is regarded by many as the most important geological site in the world. It is certainly visited by many thousands each year.

the Silurian were upended when the Iapetus Ocean closed around 420 million years ago, creating a new continent. Rivers and streams later criss-crossed this new barren land and laid down sandstones and conglomerates, building new layers of rock on the ancient surface. The resultant juxtaposition of steeply dipping rocks with more shallow-lying layers on top is known as an unconformity. Siccar Point is the first place in the world where this complex geological relationship was correctly interpreted.

Hutton had previously seen similar arrangements of strata on Arran and also at Jedburgh, but this is where the penny dropped and he understood the vast timescales involved. Hutton thought he was looking into the 'abyss of time', as he worked out this feature would have taken many millennia, or indeed millions of years, to form.

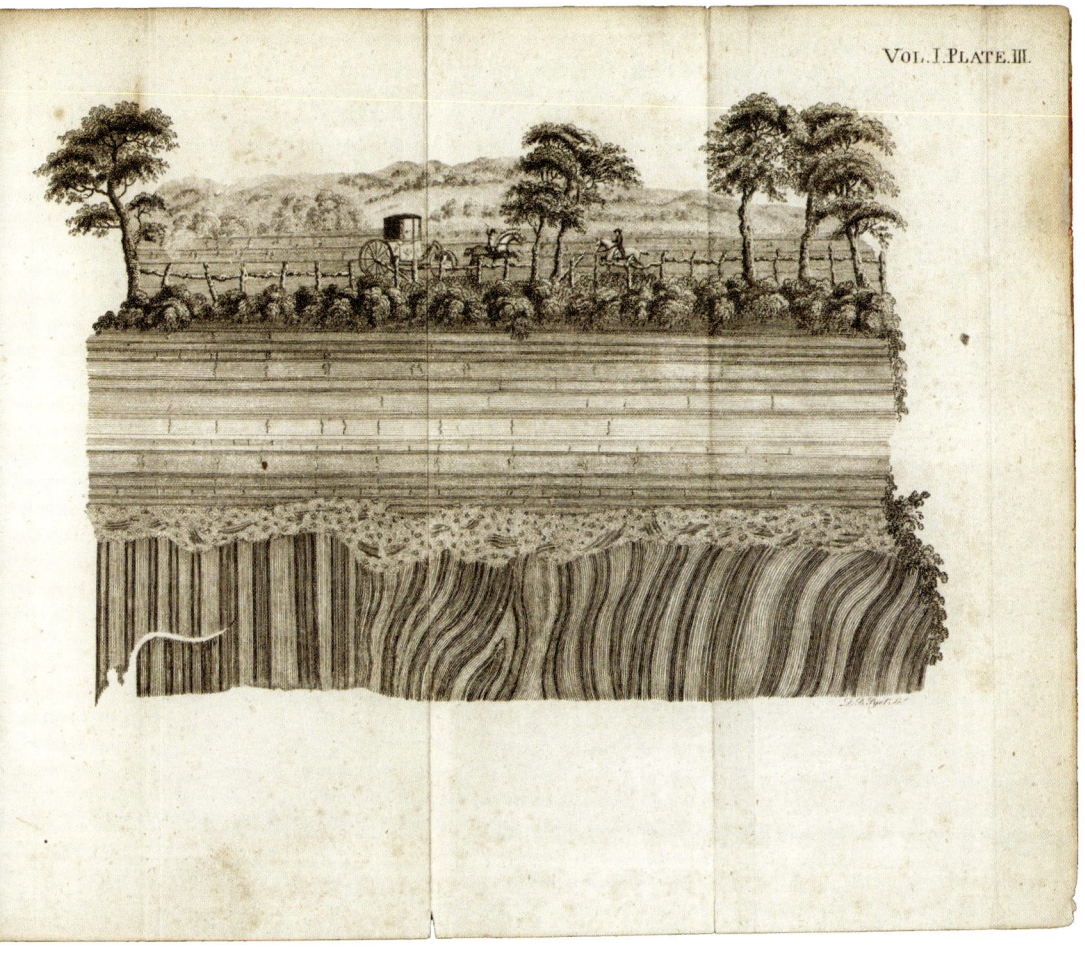

Allars Mill, near Jedburgh, was sketched by Hutton's companion, John Clerk of Eldin in 1787, and clearly shows the angular relationship between the Silurian rocks below and the flat-lying Old Red Sandstones above.

6
Volcanic interludes

St Abb's Head on the Berwickshire coast is comprised of lavas and volcanic ash that were erupted around 400 million years ago.

The Old Red Sandstone continent remained an unstable place for many millions of years. It was racked by seismic events that gave rise to tears or faults in the Earth's surface and regular outpourings of molten rock from a variety of volcanoes. The biggest of these we now share with Northumberland: the Cheviot volcano. Today, it rises to a height of just over 800 metres, comprised of a central core of granite with a lava field that is circular in plan. It is perhaps the most imposing volcano to erupt in the area. The lavas are largely classified as andesites, a rock type named from the Andes in South America. Evidence of pyroclastic flows, similar in nature to those that erupted from Vesuvius

and buried the town of Pompeii in AD 79, have also been identified. The lava and associated ash deposits were disgorged from the volcano during many violent eruptions. These events took place some 400 million years ago, when the landscape was little more than a parched desert, with little evidence of life.

Volcanoes were also active in the area we now recognise as St Abb's Head and around Eyemouth. Cones of molten rock emerged through saturated ground. The water through which the lava passed caused the volcanic activity to be more explosive than it otherwise would have been.

Volcanic activity continued into the next geological era – the Carboniferous period. The Eildon Hills, near Melrose, are the most obvious landscape feature that dates from this time. The three hills of the Eildons – Trimontium, as named by the Romans – are a dominant presence. And while they are shaped like ancient volcanoes, that impression is misleading. Indeed, they are comprised of ancient volcanic rock, but their shape is the result of a more recent event – the Ice Age.

Around 350 million years ago, thick layers of lava were erupted from a number of vents that tapped a deep source within the Earth's

The Eildon Hills are emblematic of the Scottish Borders. Sir Walter Scott, who lived at nearby Abbotsford, loved this view. It's now known as Scott's View. The River Tweed meanders lazily in front of these iconic hills.

crust that formed the bedrock of the Eildon Hills. The molten rock flowed across the landscape to form a series of thick volcanic layers. These rocks were resistant to the erosive force of the ice, so were left as upstanding peaks in a landscape that was otherwise ground flat during the Ice Age.

Elsewhere in the Scottish Border country, there is extensive evidence of volcanic activity. Around Kelso, the landscape is studded with ancient volcanic vents that spewed lava across the landscape as it existed in Carboniferous times. Around 20 of these circular plugs have been identified. In addition, thick lava flows surround Kelso in a horseshoe shape.

Similar-sized circular volcanic plugs and necks are clustered around Langholm in the south-west of the area that date from the same period. A series of lava flows, with an aggregate thickness of 90 metres, lies between Langholm and Dumfries, near Birrenswark.

Smailholm Tower stands on one of the 12 lava flows that make up the Kelso Traps, a series of lava flows erupted during Lower Carboniferous times.

7
Meandering rivers and shallow seas

As with every significant geological event, the volcanic disturbances described in the previous chapter are related to movements of the Earth's tectonic plates. We now know that plate movements to the south caused the Earth's crust to stretch in this area. This repositioning of the world order allowed molten rocks to punch through to the surface. The Earth movements responsible for this tension took place across France and Spain over a prolonged period, sending shock waves north.

Rivers and streams flowed from the upland areas that we now recognise as the Southern Uplands across this newly formed flood plain environment. The area was also periodically flooded by the sea.

The lakes and streams were home to many different types of fish,

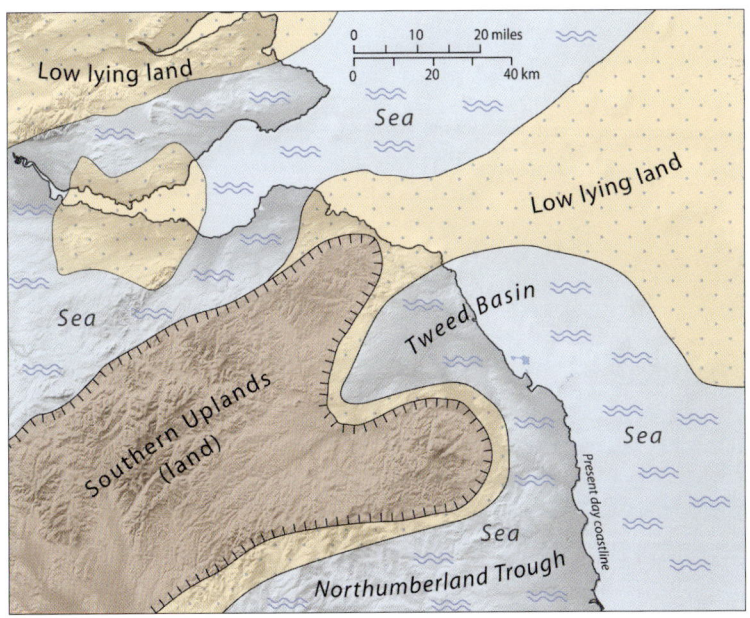

Strata were crumped when continents collided, forming a ridge of high ground. We recognise this area as the Southern Uplands today. To the east and south lay lower-lying land where new sedimentary layers accumulated. The Tweed Basin was initially a floodplain where meandering rivers criss-crossed the landscape and lakes also developed. Later, this area was flooded by the sea, as the ocean levels rose. These complex environmental changes are reflected in the geological record.

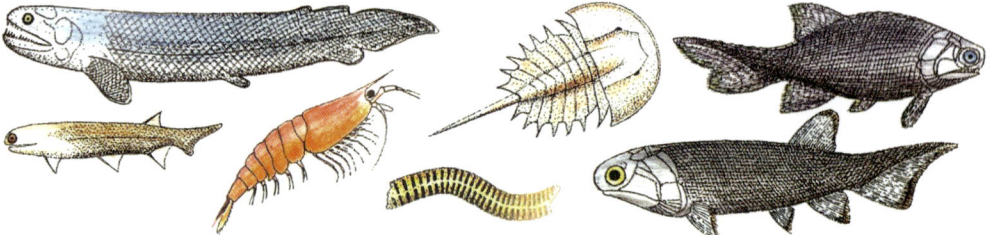

Reconstructions of the fish and arthropods found at Foulden.

king crabs, shrimps and worms. Foulden Burn, which lies just to the west of Berwick-upon-Tweed, has cut through the ancient sandstone laid down in Lower Carboniferous times. The site has been studied since 1910, when a 19-year-old boy called Thomas Ovens collected many fossil specimens from the area. Of particular interest are the 12 or more different species of fossil fishes.

A recent re-investigation of two sites close to Foulden Burn has turned up even more exciting finds. Study of the emergence of animals from water to land has always been a particular focus for palaeontologists. A gap in the fossil record, between 360 and 345 million years, was identified by American fossil expert Alfred Romer (1894–1973). It was during this period that the emergence of animals from an entirely aquatic lifestyle to that of an amphibian, living in both water and on land, was thought to have taken place. This period became known as the Romer Gap. Amazingly, the remains of a primitive amphibian, dated at around 350 million years old, have been found at two sites in this area. These sites fill the Romer Gap and are considered to be of international importance, as they provide this vital evidence of the transition between life in the water and on land. Apart from these sites in Scotland, only at one other location, known as Blue Beach in Nova Scotia, have comparable life forms, of the same age, been discovered.

This illustration imagines the emergence of primitive tetrapods from an aquatic environment to a life on land. This is a hugely important moment in the development of life on Earth.

8
Coal swamps

In Carboniferous times, this land was located on the equator. Evidence for the presence of tropical rainforests dating from this period is no surprise. Steamy swamps supporting exotic species of horsetails and clubmosses gave rise to a form of black gold – coal. When the vegetation died back, it accumulated as layers of organic matter on the forest floor. Sea levels oscillated considerably at this time, so incursions by the sea were common. Sand and mud were laid down on top of the dead vegetation and exerted pressure as layers built up. Seams of coal were thus created. This process was repeated many times. We call this the coal cycle.

Today, the extent of these coal-bearing deposits is very limited, mainly around the town of Sanquhar. Their reach was probably far greater during Carboniferous times, but was much reduced by later erosion. This economically valuable resource lay undiscovered until the industrial revolution some 300 million years later. Mining operations were initiated by the Duke of Queensbury in the late 1700s and continued until the 1980s.

Tropical rainforest environments were widely distributed across the area during Carboniferous times. Their mark on the landscape is represented by a productive coalfield that was actively mined for almost two centuries.

9
Desert storm

Driven by the irresistible forces of tectonic plate movements, the land that would become Southern Scotland was transported further north. In Triassic and Permian times, the land lay just to the north of the Equator at a latitude similar to that of the Sahara Desert today.

The next chapter in our geological story is one of barren deserts, almost bereft of life, swept by ferocious winds. The desert would have, in all probability, extended across the country, but later erosion has left us only a few remnants – around Stranraer, Dumfries, Thornhill and Lochmaben (shown as Permian and Triassic rocks on the geological map, p. 10). The desert environment consisted of towering dunes that were driven across the landscape by powerful winds. Impermanent and constantly shifting, the dunes were impressive landscape

The Scottish Portrait Gallery is one of the most magnificent buildings to grace Edinburgh's New Town. It is built from Permian sandstone from the Corsehill Quarry near Annan.

features at 250 metres in height and some 3 kilometres in length. Although limited in extent, these deposits are very thick, at around 1,500 metres or so. Preserved within some of the sediments' layers are footprints of animals that are now long extinct. Four different species of small reptile-like creatures have been identified so far. Between the dunes, there is evidence for the existence of seasonal streams created by occasional torrential downpours. These waterways would have been exactly like the modern wadis and alluvial fans found in present-day deserts.

These ancient dunes, now red sandstones, have proved to be an ideal building material. Locharbriggs Quarry near Dumfries is the most productive extraction site still working. Peak production was in the late 1800s, when 200,000 tons were quarried annually. Many of the Victorian terraces in Edinburgh and Glasgow were constructed from this material. Cargoes of the red sandstone were also exported to America.

After the desert storm of the Permian came a quieter period. In the Triassic period, around 250 million years ago, wide, slow-moving rivers meandered across the arid landscape. Structures preserved in the rocks from this time indicate that these desert rivers occasionally dried up completely. The sedimentary layers laid down during this time proved to be excellent stone for building and have been used locally.

The following Jurassic and Cretaceous periods left no imprint on Southern Scotland. We look to the Palaeogene period for the next geological event that left a mark here. From Permian times onwards, Scotland continued to drift northwards to higher latitudes. For many millions of years, Scotland lay within the supercontinent of Pangaea. With another substantial continental re-arrangement, this landmass split asunder and the North Atlantic Ocean began to open around 65 million years ago. From this point onwards, Europe and North America went their separate ways. This pulling apart of the Earth's crust led to this upper layer becoming significantly thinner and, as a result, huge volcanoes punched their way to the surface. Massive eruptions formed the land that became Skye, Rum, the Ardnamurchan area, Mull and Arran. It is the Mull volcano that left a mark on Southern Scotland. Thin ribbons of molten rock were shot out of the Mull pressure cooker in a southeasterly direction and travelled for 200 miles across Southern Scotland to north-east England and out into what is now the North Sea. They are little more than a few metres across, but impressive, nevertheless, as an indicator of the immense volcanic forces at work.

10
Mineral treasures

At Wanlockhead, near Abington, there is evidence of mining that dates from the middle of the eighteenth century. There is still much to see, and the presence of the Museum of Lead Mining on site enhances the visitors' experience. The principal ores mined at this site are galena (lead sulphide) and sphalerite (zinc iron sulphide). Between 1750 and 1958, the date its commercial operation ceased, some 400,000 tonnes of lead and 10,000 tonnes of zinc were recovered. A smelter was operated on site. Some of the ancient galleries, created by the miners as they followed the mineral veins underground, are still intact.

The commercial value of the mine was considerably enhanced by the presence of gold and silver. Some 25 tonnes of silver were recovered and the sediments carried in nearby streams are rich in gold. Gold-panning championships are still held annually in Wanlockhead. In 2015, a gold nugget valued at £10,000 was found in a stream at this location. This alluvial gold is also of great historic importance. After the discovery of precious metals at this site in the early sixteenth century, the gold in particular was used in significant quantities for the creation of royal crowns and jewellery, and also for the minting of coins. The ceremonial mace, which was presented by HM The Queen at the opening of the Scottish Parliament in 1999 was, in part, made from gold panned at Wanlockhead.

A specimen of galena.

11
The Ice Age and beyond

Rapidly falling temperatures ushered in an Ice Age about 2.6 million years ago. This was a worldwide event, affecting the countries in the highest and lowest latitudes particularly. As temperatures fell, tongues of ice reached southwards from the north polar region, to the extent that the whole of Scotland was submerged. Over the next 2 million years the mark the 2-kilometre-thick cover of ice and snow made on the landscape was profound.

The Ice Age was not a single event but a series of advances and retreats of the ice fields in response to the changing climate. When temperatures were at their coldest, the whole area would have been completely shrouded by a freezing blanket. Between these times, when the cover of ice reached its maximum extent, warmer, more forgiving climates prevailed. These are known as inter-glacials. Since the last of the ice melted around 11,000 years ago, we have been in an inter-glacial period, but in all likelihood the ice will return – the current estimate being in about 55,000 years' time. This figure takes into account the currently predicted effects of climate change.

This see-saw of global temperatures is driven by some factors over which we have absolutely no control. Our orbit around the Sun varies from circular to elliptical. It takes around 100,000 years for the cycle

This chart shows the wildly oscillating temperatures experienced during the last 3 million years. Warm periods are swiftly followed by cold, when the glaciers build once again. Climate change isn't a recent phenomenon; it's been going on for millions of years. What is new is the fact that since the Industrial Revolution the greenhouse gasses released into the atmosphere as a result of burning fossil fuels are exacerbating the current steep rise in global temperatures. Damaging rises in sea level will result, unless urgent action is taken.

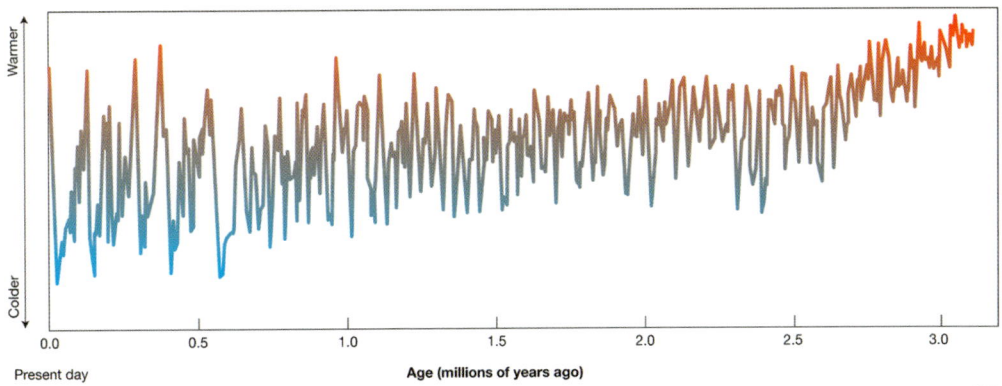

33

Ice and snow accumulated to the point where very little of the landscape we are familiar with today was visible.

between a circular orbit to an ellipse to be completed. When the Earth is furthest from the Sun, the temperature drops and ice builds up in the polar regions and beyond. But when the planet is in a completely circular orbit, the climate warms up and the ice cover shrinks. There have been periods in the geological past when the poles have been entirely ice-free. The Gulf Stream, a current of warm water that flows across the Atlantic from the Gulf of Mexico, has also had an important moderating effect on climate through recent times.

The ice sheet was not static. Glaciers move in response to gravity.

The Devil's Beef Tub is an extraordinary landscape feature carved out by the ice.

As the frozen rivers of ice travelled from higher ground, they carved and moulded the bedrock to the familiar shapes we see today. Great glacial troughs were gouged deep into the bedrock and the higher ground was rounded into soft pillow-like shapes. Detailed study of the area reveals that in the mountains around Loch Doon, in the west, a particularly thick accumulation of ice and glaciers radiated from that point. We know this because of scratches made by the glaciers on rock surfaces, known as striations, and also the general pattern of glacial erosion that fanned out from the Loch Doon area.

Large boulders carried by the ice were dumped randomly across the countryside as the ice melted. These large lumps of rock were torn from the bedrock and transported by the moving ice to a location many kilometres away. They are known as glacial erratics. Some of the blocks of granite have been transported south into northern England. This photo also shows the erosive power of the ice in creating this pavement carved into the underlying granite. The location of this fine example is at the Devil's Bowling Green near Craignaw in the Galloway Forest Park.

The Devil's Beef Tub near Moffat is an excellent example of the erosive power of the ice. It has carved a deep depression in the bedrock, creating one of the most impressive landscape features in the area.

As the climate began to warm towards the end of the last glacial advance, about 11,000 years ago, the ice sheets melted away in the milder conditions. The burden of large rocks, sand and mud carried by the glaciers was dumped and formed a series of interesting landforms that are characteristic of this de-glaciation. The volume of water liberated by this ice-melting process was colossal. What are known as meltwater channels, deep grooves in the landscape cut by these raging torrents of water, criss-cross the area.

As the ice melted, rivers occasionally formed between bedrock and the overlying ice. When the current of these flowing waters slackened, rocks, sands and muds carried by the rivers were lain down along their winding course. One of the best examples of this landform, called an esker, is on Greenlaw Moor (see Places to visit, p. 45).

THE ICE AGE AND BEYOND

These hummocks are known as moraines. They are piles of boulders, sands and mud dropped by the moving ice and then shaped into these distinctive forms. These glacial deposits are near Loch Skene, close to Moffat.

The winding river course under the ice is traced by the Greenlaw Moor esker.

12
The coast

The east and west coasts of Southern Scotland present something of a contrast. The east is made of hard, resistant rocks that have held back the storm waters of the North Sea since the ice melted 11,000 years go. In contrast, the west, particularly around the Solway coast, is softer, wrapped in saltmarsh and sandy beaches.

St Abb's Head, 65 km east of Edinburgh, is a magnificent rugged coastline. Gullies, fault-guided inlets, geos, sea stacks and coves cut into the lavas and ashes of Devonian age have created a dramatic coastal landscape. The headland is very exposed to gales coming from the north and north-east, so the cliffs are regularly hammered by lively

The rugged coastline at St Abb's Head.

sea conditions. They are some of the finest cliffs to be found anywhere in the UK, so well worth a visit.

The Solway coastline is fringed by a huge saltmarsh. This is a coastal ecosystem of salt-tolerant grasses and mudbanks that is regularly washed by the sea. It stretches almost continuously between the Scottish coast and St Bees Head on the English side, a distance of around 60 km. Because of the configuration of the estuary, there is a net build-up of sediment, so this habitat acts as an excellent coastal defence against storms coming from the Irish Sea.

Luce Sands, south of Stranraer, is a fine example of a large and complex system of beach sands and fringing dunes. It is highly dynamic, as it moves and changes form with the prevailing winds and tides. Inland from the tidal area is Torrs Warren, described as the largest complex of dunes in Southern Scotland. Its undisturbed nature is due to the fact that it is part of a military range. It is also an important host site for an internationally significant number of hen harriers, who use this area as a refuge during the winter months.

Cattle graze on the saltmarsh in the Solway Firth.

13
Human imprint on the landscape

The land that emerged after the ice melted around 11,000 years ago was barren and boulder strewn. Gradually, pioneer tree species of juniper and birch clothed the hills and valleys. Then came oak, hazel and elm. The first evidence of a human imprint on the landscape arrived around 5,000 years ago, with the cutting down of part of the

Kelso Abbey
Magnificent places of worship, like the impressive Kelso Abbey, were built in the 1100s. Similar structures were constructed at Melrose, Jedburgh and Dryburgh in the east, and Sweetheart Abbey in the south-west. Frequent attacks from unfriendly neighbours left this building in ruins. These abbeys were imposing features on a landscape scale that would have dominated the lives of many for around 400 years.

Grouse moors
Innovative management techniques are being trialled to evaluate whether heather moorlands can support economically viable driven-grouse businesses whilst also accommodating endangered birds of prey. The most advanced of these trials has taken place at Langholm, where Buccleuch Estates are working with various conservation organisations.

forest to accommodate a more settled pattern of agriculture. This change is reflected in the pollen archive recovered from local peat bogs. There is also an increase in pollen from cereal crops that indicates a thriving, albeit rudimentary, agricultural economy.

Archaeological finds throughout Southern Scotland, such as arrowheads and flint tools, confirm a human presence from these early times. This 10,000-year period, now known as the Anthropocene, charts the course of antiquity from our distant geological and glacial

Forestry
Managed forestry plantations have had a significant effect on the landscape of this area, particularly in Dumfries and Galloway. The Forestry Commission (now known as Forest and Land Scotland) was founded in 1919 to provide employment for demobbed soldiers and also to increase the tree cover throughout the country. Its widespread footprint remains to this day.

past, through archaeological evidence and artefacts, to a time when formal written historical records were kept. This new term 'Anthropocene' acknowledges the impact that human presence has had, and is having, on the planet in terms of the effects on all aspects of the environment – atmosphere, geosphere and hydrosphere.

In Southern Scotland we see the landscape being impacted in many ways by people, including the construction of buildings, forestry and wind farms, and the trialling of grouse businesses, in particular at Langholm.

Wind farms
Onshore wind projects are now economically viable and are making a significant contribution to our diminishing carbon footprint. Southern Scotland is a favoured location for wind turbines, as the population density is low over much of the area and the winds are strong.

14
Places to visit

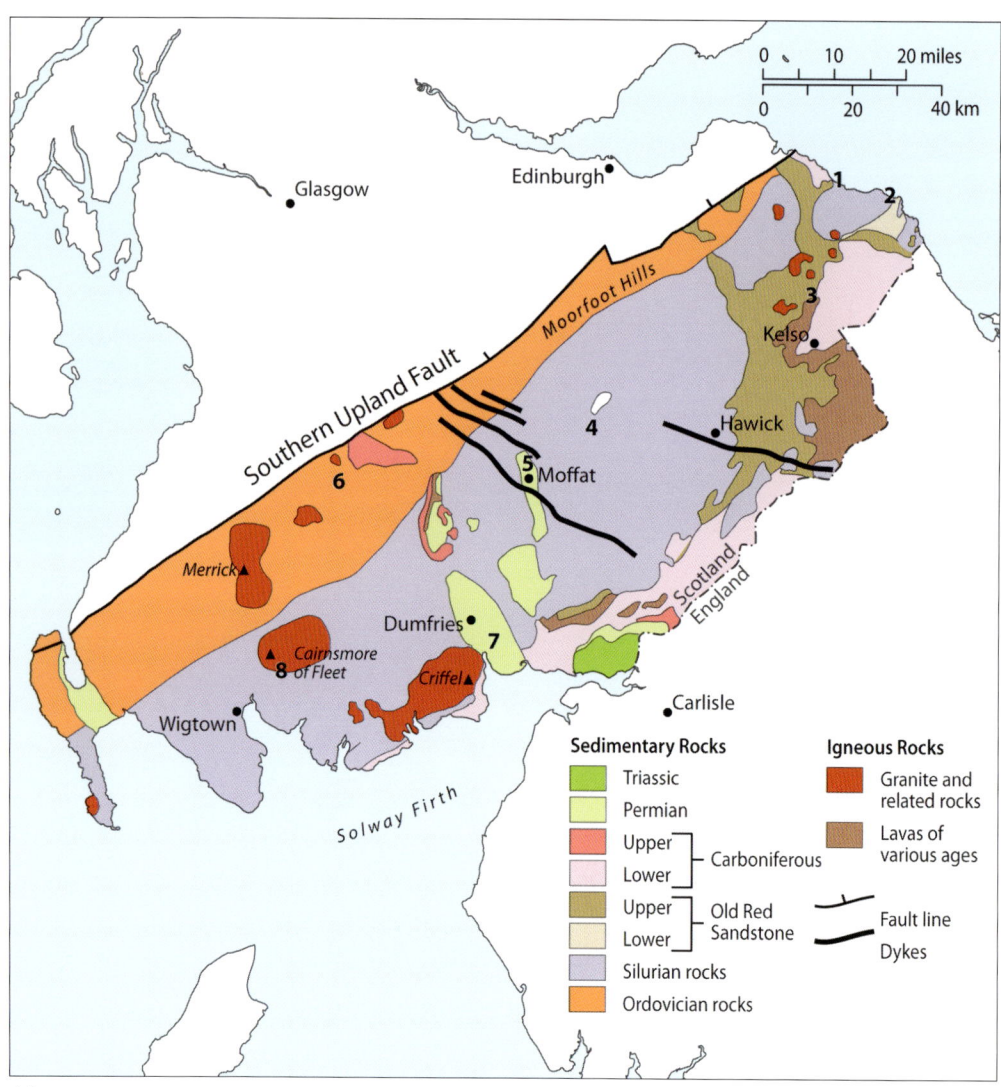

Much of the higher ground is hidden beneath a cloak of heather or, in Dumfries and Galloway in particular, commercial forestry. Many of the key geological sites are exposed along the coastline and in rock sections cut by rivers and streams. The remainder of the area is, in the Tweed Valley particularly, high quality agricultural land. This list of key sites should help those who are unfamiliar with the area. It is covered by 12 1:50,000 scale OS Landranger maps and the Bedrock Geology UK North map, published by the British Geological Survey. These maps will help you plan and execute your visit to the area.

1. Siccar Point: on the Berwickshire coast, this is regarded by many as the most important geological site in the world. It was the place where James Hutton went to test his ideas, outlined in his seminal book *Theory of the Earth*. Published in 1795, it formed the basis of the modern science of geology. Travelling south, turn off the A1 just south of Cockburnspath and follow the road to the coast. There is a small car park and signage for the benefit of visitors. The site can be viewed from the top of the cliff. There is no path to the base of the cliff, so descend at your own risk. You have been warned. But it's well worth the effort to stand where Hutton stood when he made his revelatory discoveries. The site is visited by countless domestic and foreign tourists each year, many of them geologists, to see where their science began.

The wave-washed exposures at Siccar Point were visited by James Hutton and companions in 1788. This place was the inspiration for one of the most important and influential ideas of the eighteenth century.

Grey Mare's Tail waterfall is an impressive feature, much visited by tourists to the area.

2. St Abb's Head: the geology and dramatic cliff line of St Abb's Head have already been described. There is good vehicle access and parking at the site. The lighthouse adds an extra dimension to the visit.

3. Greenlaw Moor: an area of heather moorland near the town of Greenlaw in the Scottish Borders. The interest here is two-fold. The esker is described earlier (p. 36). In addition, there are areas of wetland that support populations of pink-footed geese, which are regarded as important in the European and world context.

4. Grey Mare's Tail: a spectacular waterfall located just to the north of the A708 Moffat to Selkirk road. A stream, fed by the adjacent Loch Skeen, tumbles some 60 metres down the hillside to the valley below. There is an adjacent car park and footpath leading to the base of the waterfall. The site is managed by the National Trust for Scotland.

5. Devil's Beef Tub: there are strong literary associations with Sir Walter Scott here. In his novel *The Red Gauntlet*, he describes this deep depression in the land surface thus: 'It looks as if four hills were laying their heads together to shut out daylight from the dark hollow spaces between them. A damned deep, black, blackguard-looking abyss of a hole it is.' It's an 11 km hike northwards from Moffat along the Annandale Way, following the River Annan, but it's worth the trek.

Caerlaverock Wetland Centre.

6. The Museum of Lead Mining, near Abington: this is a great day out. It has a museum, shop and a tearoom and offers short courses in gold-panning. It is also possible to take an accompanied trip underground to see the galleries created when the mine was first worked more than 200 years ago.

7. Caerlaverock Wetland Centre, on the north shore of the Solway Firth, south of Dumfries: an excellent day out. Managed by the Wildfowl and Wetland Trust, it offers a range of features of interest. The saltmarshes of the Solway Firth are of national landform interest in terms of how this system has evolved over time in response to changing sea levels and channel migration patterns. The bird life is spectacular, and there are 20 different hides and trails to get you closer to nature. There is also a café and a shop.

8. Cairnsmore of Fleet: this national nature reserve is underlain by a large granite hill that supports a diverse range of habitats, including uplands, heather moorland and blanket bog. It is also the home to bird species including wheatear, skylark, curlew, and black and red grouse. If you are lucky, you will spot a golden eagle! On site, there are trails, a visitor centre and ample car parking. It's also a great place to connect bedrock geology with landscape. The granite was resistant to erosion during the last Ice Age so, today, the Cairnsmore of Fleet granite towers above the surrounding countryside.

Acknowledgements and picture credits

Thanks are due to Professor Stuart Monro OBE FRSE and Moira McKirdy MBE for their comment and suggestions on the various drafts of this book. I also thank Debs Warner, Mairi Sutherland, Andrew Simmons and Hugh Andrew from Birlinn Ltd for their support and direction. Mark Blackadder's book design is up to his usual very high standard. I thank the James Hutton Foundation for their financial support in the publication of this book. Scottish Natural Heritage, in association with the British Geological Survey, published the *Landscape Fashioned by Geology* series that was the precursor to the new *Landscapes in Stone* titles. I thank them both for their permission to use some of the original artwork and photography in this book. David McAdam, Phil Stone and Andrew McMillian are the BGS authors who wrote the original texts. I dedicate this book to Alistair Moffat. We grew up together on the same mean streets of Kelso in the beautiful Scottish Borders. Alistair has so many accomplishments to his name – Edinburgh Festival Fringe director, STV director and programme maker, 'Scotland's historian', founder of the Borders Book Festival, former Rector of St Andrews University and, most recently, champion of the Great Tapestry of Scotland and its permanent home in Galashiels.

Picture credits

2–3 Lorne Gill/SNH; 6 Lorne Gill/SNH; 10 Helen Stirling (map); 12 (upper) drawn by Jim Lewis, (lower) drawn by Robert Nelmes; 13 Craig Ellery; 14 Clare Hewitt; 15 Lorne Gill/SNH; 16 © National Museums Scotland; 19 Lorne Gill/SNH; 20 (upper) Helen Stirling (diagram), (lower) Lorne Gill/SNH; 21 drawn by Jim Lewis; 22 Alan McKirdy; 23 Sir Robert Clerk of Penicuik; 24 Lorne Gill/SNH; 25 Ulmus Media/Shutterstock; 26 Bill McKelvie/Shutterstock; 27 Helen Stirling (diagram); 28 Professor Euan Clarkson; 29 Elizabeth Pickett/North Pennines AONB Partnership; 30 Claudine Van Massenhove/Shutterstock; 32 Albert Russ/Shutterstock; 34 John Gordon; 35 Shirley Kilpatrick/Alamy Stock Photo; 36 Lorne Gill/SNH; 37 (upper) Lorne Gill/SNH, (lower) Patricia and Angus Macdonald/Aerographica; 38 Mark Pink/Alamy Stock Photo; 39 Richard P Long/Shutterstock; 40 Jule-Berin/Shutterstock; 41 Lorne Gill/SNH; 42 Lorne Gill/SNH; 43 Lorne Gill/SNH; 44 Helen Stirling (map); 45 Lorne Gill/SNH; 46 Martin Valigursky/Shutterstock; 47 Lorne Gill/SNH.